The Mighty O

by

Art Giberson

The Mighty O

By
Art Giberson

Second Edition

ISBN-13: 978-0-9845777-5-0

Technical Review Editor - Nelson O. Ottenhausen
Cover Design – Art Giberson
Managing Editor - Dari Bradley

The events and places in this publication are historically true and accurate as written by Art Giberson in his daily duties while serving a in the United States Navy as a naval photographer.

Published by Patriot Media, Inc.
Publishing America's Patriots
P.O. Box 5414
Niceville, FL 32578
United States of America
www.patriotmediainc.com

IN MEMORY OF:

Vice Admiral Jack Fetterman, USN, (Ret.)

ACKNOWLEDGEMENTS

Thank you to those who assisted with the development of this book, especially my son, Tony, for suggesting that I write it.

Special thanks to my wife, Jean for her undying support.

A heartfelt thanks is extended to my friends at Patriot Media, Inc., Gulf Coast Authors and to Terry and Helen Spears for their encouragement and faith in this project.

INTRODUCTION

The aircraft carrier *Oriskany*, "*The Mighty O*" served the United States in war and peace for more than a quarter century. *Oriskany* launched its first air strikes against a hostile foe during the Korean War and was credited with shooting down two North Korean MiG-15s. Although *Oriskany* combat action was limited during the Korean War it was still enough to establish *Oriskany* as one of the Navy's most formidable combatants of the post-World War II era.

Within days of the America forces landing at Da Nang, South Vietnam, *Oriskany* aircraft were flying combat missions against the Viet Cong. These early combat operations earned *The Mighty O* and embarked Carrier Wing 16 the Navy Unit Commendation for exceptionally meritorious service. Between May 10th and December 6th 1965, *Oriskany* aircraft flew more than 12,000 combat sorties and delivered nearly 10,000 tons of ordnance against enemy forces. By the time America's involvement in Vietnam ended in 1973, *The Mighty O* had served the equivalent of four full years in combat.

For her wartime service *Oriskany* received two battle stars for Korean service and five for Vietnamese service.

The Mighty O

Chapter 1

From World War II to the present, American aircraft carriers, acting as forward deployed combat-ready units, have traditionally been the first military forces to show the flag during a potential hostile situation. One of those carriers, the USS *Oriskany* (CVA-34), affectionately known as "***The Mighty O***", was one of those front line units and saw combat action in two wars—Korea and Vietnam.

Oriskany, an Essex-class carrier, was launched on the Navy's 175th birthday October 13, 1945. With World War II over and a drastic downsizing of the American military, construction of warships came to an abrupt halt. That included *Oriskany*, despite the fact that the ship was better than 85 percent complete. Unlike many other Navy vessels under construction and sold for scrap after the war, construction of CV-34 resumed in August 1947, but with substantial design changes which basically made *Oriskany* a one-of-kind aircraft carrier.

Oriskany was the prototype for a modernization program called SCB-27, which allowed the ship to handle the new, more powerful and heavier generation of carrier aircraft developed after the war. The redesigned *Oriskany* had a massive reinforced flight deck, capable of handling the largest aircraft in the fleet at the time. Stronger elevators, more powerful hydraulic catapults, and new arresting gear were also installed. The island structure was rebuilt, antiaircraft turrets, which were common on the older Essex-class carriers, were removed and an armor belt, originally installed to protect the carrier from torpedoes, was removed and her hull was widened.

When she was finally commissioned on September 25, 1950, *Oriskany*, for all intents and purposes, was in a class all by itself. Unofficially dubbed an *Oriskany*-class carrier, *The Mighty O* was capable of operating aircraft weighing upwards of 50,000 pounds. Following formal commissioning ceremonies *Oriskany* put to sea almost immediately for her first homeport, Naval Air Station Quonset Point, Rhode Island. During the first nine months of her operational service *Oriskany* conducted extensive sea trials, training and deployment preparations. On May 15, 1951, *Oriskany* departed Quonset Point for the Mediterranean, marking the first of 17 overseas deployments she would make over the next 25 years.

Returning to the United States in October, *Oriskany*, after five months at sea, was ready for an overhaul. The overhaul, performed at the New York Navy Yard, included the installation of a new flight deck, bridge, and the installation of an automatic steering system—the first such system to be installed on an aircraft carrier. The complex "auto-pilot" made extensive use of gyroscopes and operated from a single pilot-house console. This was the first of many *Oriskany* "firsts."

The yard period over, *Oriskany* departed the shipyard May 15, 1952, for Norfolk, Virginia, where she took on ammunition then got underway to join the Pacific Fleet, at San

Diego, California. Making port calls at Guantanamo Bay, Cuba, and Rio de Janeiro, Brazil, *Oriskany* established another first when she became the first aircraft carrier to travel from the Atlantic to the Pacific by sailing around Cape Horn on the Southern tip of South America.

Hardly before the crew of *The Mighty O* had time to get acquainted with their new homeport, *Oriskany*, with Carrier Air Group 12 embarked, sailed for Korea on September 15, 1952 for her first combat deployment—a deployment that established *Oriskany* as a super combatant of the post-World War II era.

By the time *Oriskany* arrived at Yokosuka, Japan, to join Fast Carrier Task Force-77 her designation had been changed from CV (carrier) to attack carrier (CVA). Within days of joining Task Force 77, the newly reclassified attack carrier launched her first combat air strikes against an enemy force.

Carrier Air Group Twelve (CAG-12) struck hard and furious with bombing and strafing attacks against North Korean supply lines, troop encampments and coordinated bombing missions and surface gun strikes along the coast.

In her first combat deployment *Oriskany*'s air group flew more than 7,000 sorties, dropped 4,600 tons of bombs, fired more than a million rounds of ammunition, and shot down two Soviet built MiG-15s.

After her initial and successful combat engagement, *Oriskany* sailed for Japan for a much needed rest and brief upkeep period. Departing Japan on March 1, 1953, *Oriskany* returned to the combat zone and resumed flying aerial support missions for another 10 days before finally departing for Hong Kong and some rest and relaxation for the crew. As she dropped anchor in Hong Kong Harbor, there was little doubt that the last of the famed Essex-class carriers had earned its nickname —*The Mighty O*.

U.S. Navy Photo

Ready for action: *Her flight deck bulging with F9F-5 Panthers FU-4 Corsairs and A-1 Skyraiders,* Oriskany *replenishes at Yokosuka, Japan, prior to departing for UN support operations off the coast of Korea.*

Oriskany didn't escape the war zone unscathed however. On March 6, 1953, a bomb from an F4U Corsair, returning from a combat mission, dislodged when the plane landed and exploded. Fire quickly spread to the hangar deck.

Fortunately the fire was isolated in the hangar bay and extinguished, but not before killing two Sailors and injuring 15 others. One of the Sailors killed was a photographer who filmed the bomb striking the deck just seconds before it exploded. This was the first of *The Mighty O*'s encounters with fire at sea, but regrettably it wouldn't be the last.

(Official U.S. Navy foto via United Press Telephoto)

Death Comes to the Cameraman. Stills from film in movie camera operated by Navy fotog Thomas L. McGraw Jr. of Theresa, N. Y., show approach of death in form of live aerial bomb. After breaking loose from fighter plane (left), which was landing on the carrier Oriskany off Korea, bomb bounced along flight deck to explode with blinding flash directly in front of camera. McGraw and another man were killed by the blast; 15 other members of the Oriskany's crew were injured. See fotos below.

Carrier Oriskany steams up East River during visit to New York in November, 1951.

The late Thomas L. McGraw Jr.

Newspaper clipping courtesy Oriskany Museum, Oriskany, N.Y.

Much to the crew's regret *Oriskany's* return to the States on May18, 1953 was short lived. After a mere four-month turn around, much of which was spent at sea preparing for the next deployment, *Oriskany* and her combat-weary crew, again departed for the Western Pacific; her third overseas deployment in as many years.

This time, however, instead of attacking North Korean fortifications, she sailed into the annals of motion picture stardom as the fictitious carrier, *Savo*; in the 1954 box office smash *The Bridges at Toko Ri*, with William Holden and Mickey Rooney.

One of the pilots serving aboard *Oriskany* at the time was future astronaut, Alan B. Shepard. According to Shepard's autobiography, ***Light this Candle***, during the filming of *The Bridges at Toko Ri*, Shepard, Rooney and another aviator, rumored to have been a former squadron mate of Shepard's, would quietly retire to Shepard's stateroom at the end of the day for alcoholic refreshments, which of course was, and are, strictly forbidden aboard U.S. Navy ships. But then, the first American to fly in space was noted for bending the rules.

With war and motion picture stardom behind her, *Oriskany* settled into a peacetime routine. That routine was temporarily shattered on March 2, 1954, when an F2H-3 Banshee returning from a training flight, struck the after ramp of *Oriskany*'s flight deck and broke in half.

Skidding across the deck in a ball of flame, the Banshee smashed into the island. Miraculously the pilot, reportedly Shepard's stateroom drinking buddy, John Mitchell, survived and the flames were extinguished before they could spread to nearby aircraft. The Banshee accident, in effect, ended what had been a scheduled six- month deployment and *Oriskany* returned to San Diego.

After the damage to her island structure was repaired, *The Mighty O* again put to sea for routine operations and once again the carrier was thrust into the Hollywood spotlight, this time as the floating star of James Michener's "The Men of the Fighting Lady."

With the completion of the film, *Oriskany* entered the shipyard for a regularly scheduled ship's maintenance period which lasted until February 1955.

Exiting the shipyard in March, the gallant warrior once more deployed to the Western Pacific for a seven-month deployment. During this deployment *Oriskany* and her air group, Carrier Air Group Nineteen (CAG-19) won the prestigious Commander Naval Air Pacific Battle Efficiency "E" Award.

Naval aviation had changed considerably by the mid-1950s. Faster and heavier aircraft were entering the fleet requiring radical changes in the design and capabilities of aircraft carriers. New carriers such as the USS Forrestal (CVA-59), commissioned October 1, 1955, were coming on line and many of the older Essex-class carriers were being taken out of service.

Since the end of the Korean War, *Oriskany* had regularly deployed to the Western Pacific for back-to-back tours of duty with the Seventh Fleet and was long overdue for a major modernization. Her very survival as a front line, combat capable, aircraft carrier demanded it.

But before modernization could occur one final deployment was in the works—her last as a straight deck carrier. Upon returning from her fourth deployment in June 1956 preparations were made for *Oriskany*'s final modernization.

Destined to become a super carrier, *Oriskany* was the last Essex- class carrier to be modernized and the only one of the 24

Essex-class carriers to receive steam catapults. Although Essex-class carriers would continue as part of both the Atlantic and Pacific Fleets for many years, their role had changed and all were converted to antisubmarine (CVS) carriers.

When *Oriskany* once again unfurled her commissioning pennant and returned to sea in 1959, she was two-and-half-tons heavier, with an angled flight deck, steam catapults, an enclosed hurricane bow, a second deck-edge elevator and a vastly improved Combat Information Center. The true legacy of *The Mighty O* was about to begin.

After a shakedown period and underway training, *Oriskany*, with a new air group (Carrier Air Group-14) and high performance aircraft aboard, sailed out of San Diego on May 14, 1960 for a seven month Far Eastern deployment. When *Oriskany*, by now a WestPac veteran, arrived at the various Western Pacific and Far Eastern ports of call, with Carrier Air Group-14 (Carrier Air Groups were later renamed, Carrier Air Wings—abbreviated CVW) deployed aboard, everyone quickly noticed that *The Mighty O* was a vastly different warship, from the carrier WestPac sailors and locals had grown accustomed to seeing during her previous five deployments.

Anxious to prove herself as a "super" carrier, *The Mighty O* completed her first post-modernization deployment without a single mishap and recorded her 5,185th arrested landing, proving beyond a shadow of a doubt that this former Essex-class carrier was worthy of its nickname and capable of holding its own with any aircraft carrier in the United States Navy.

During the Vietnam War, *Oriskany* was on the gun-line for up to nine months at a time for every year of the war (1965-1975) except one. Translated into actual combat service, *Oriskany* and her air groups, during the Vietnam War, spent more than two-and-a-half years in combat. When compared to World War II, which lasted for four years, *The Mighty O*'s actual combat record is viewed in an entirely different light, easily placing *Oriskany* at the top of the list of wartime aircraft carriers.

US Navy Photos

Ready to join the Fleet: Oriskany *(right) as she appeared after being placed in commission in 1950, en route to the Jacksonville, Florida, operating area for carrier qualification in preparations for her first overseas deployment.*

Super Essex-class: *After her final moderation in 1959,* "The Mighty O" *showed little resemblance to her sister Essex-class carriers. The highly improved* Oriskany *sported a new angled flight deck, aft deck edge elevator, new more powerful steam catapults and enclosed hurricane bow.*

The wooden flight deck planking was replaced with aluminum planking.

Chapter 2

Most Americans consider the March 1965 Marine landing at Da Nang, South Vietnam, as the actual commencement of American involvement in Southeast Asia; and while that was the first major insertion of ground troops in Vietnam, Oriskany's involvement actually commenced nearly two years before the first American troops waded ashore at Da Nang.

The *Mighty O* was midway through a nine-month deployment In August 1963, when she was dispatched to the South China Sea, near the coast of South Vietnam, following the overthrow of South Vietnam President Ngo Dinh Diem. Though no shots were fired and no bombs dropped, those few weeks of steaming off Vietnam marked the beginning of a decade of Vietnam deployments for *Oriskany* and her air group.

U.S. Navy Photo

A VA-23 A4 Skyhawk takes its place in the flight pattern for recovery aboard Oriskany *during flight operations in the Tokin Gulf.*

Retuning to San Diego in March 1964, *Oriskany* entered Puget Sound Naval Shipyard for a routine overhaul. Following the overhaul, she again put to sea for refresher training and carrier qualifications with Carrier Air Wing-16. Carrier qualifications this time around however, were far from the routine matter of getting the crew re-accustomed to the rigors of flight deck operations. This time out, carrier qualifications included testing the Navy's new airborne early warning aircraft, the E-2 Hawkeye.

U.S. Navy Photo

Oriskany *flight deck personnel watch as an E-2 Hawkeye, a Navy airborne early warning aircraft practices touch-and-go landings aboard* The Mighty O.

Several versions of the E-2 would be introduced over the next several years and serve as airborne aircraft control centers for both American and Allied forces. The newer faster Grumman E2 Hawkeye eventfully replaced the E-1

Tracer as the far-ranging "eyes and ears" for the Atlantic and Pacific Fleets. At the completion of carrier qualifications and testing the E-2 Hawkeye for carrier use, *Oriskany* departed San Diego April 5, 1965, for what would become her first official Vietnam combat deployment.

Although *Oriskany* served as the primary testing deck for the E-2 Hawkeye, for most of her seven Vietnam deployments, the smaller but versatile E-1 Tracer, a.k.a. Willie Fudd, served as the ship's early warning aircraft and airborne traffic control center.

A Willie Fudd (right) is moved into launch position by Oriskany flight deck personnel during combat operations in Southeast Asia, 1969.

U.S. Navy Photo

Teaming with aircraft from the USS *Coral Sea* (CVA-43), *Oriskany*- based Carrier Air Wing-16, commanded by Commander James B. Stockdale, added their weight to the massive American naval strength in Southeast Asia by systematically attacking bridges and highways in North Vietnam. *The Mighty O* had arrived on station.

Between May 10 and December 16, 1965, Carrier Air Wing-16 aircraft flew more than 12,000 sorties and delivered nearly 10,000 tons of ordnance against enemy forces. *Oriskany* and her air wing's baptism of fire earned her and CAW-16 the Navy Unit Commendation for exceptionally meritorious service during combat operations.

Although well deserved, the award carried a very high price tag. Less than 90 days after arriving on station *Oriskany* recorded her first combat loss of the war when Lt.(jg) E.A. Davis, flying a prop-driven Douglas A-1H "Spad" was hit by anti-aircraft fire while on a bombing mission. The pilot ejected and landed safely but was quickly captured by the North Vietnamese.

Two weeks later, September 9, 1965, the Wing Commander, Cmdr. James Stockdale, was catapulted off the flight deck of *Oriskany* for what turned out to be his final mission over North Vietnam. As his aircraft approached the target, his plane was riddled with anti-aircraft fire. Within seconds, the engine was aflame and all hydraulic control was gone. Ejecting safely he slowly floated earthward while watching his plane slam into a rice paddy and explode in a fireball. Immediately upon touching the ground he too, was taken prisoner and held captive for nearly eight years. Upon release, the future Independent Party vice presidential candidate was awarded the Medal of Honor for his superior leadership as the senior American POW in North Vietnam.

According to the citation accompanying the Medal of Honor: 'Stockdale deliberately inflicted a near-mortal wound to his person in order to convince his captors of his willingness to give up his life rather than capitulate. He was subsequently discovered and revived by the North Vietnamese who, convinced of his indomitable spirit, abated in their employment of excessive harassment and torture toward all Prisoners of War.'

Cmdr. James B. Stockdale, Commander CVW-16, deployed aboard Oriskany, *was shot down over North Vietnam.*

Stockdale's courage and endurance continue to serve as an inspiration to American service men and women of all branches of the military.

U.S. Navy Photo

During his incarceration Cmdr. Stockdale was tortured and forced to wear vise-like heavy leg irons for two years, and spent four of the seven years as a POW, in solitary confinement—in total darkness.

Though his captors held his body prisoner, their relentless attempts to break his spirit never succeeded. Throughout his captivity, Stockdale's steadfast refusal to cooperate with the enemy kept alive the spirit of resistance in his fellow POWs. When he was released in 1973 and the world learned of his courage and endurance he instantly became an inspiration to Americans everywhere.

A 1947 graduate of the U.S. Naval Academy, Vice Adm. Stockdale passed away June 5, 200[illegible]

By the time *Oriskany* returned from her first Vietnam combat deployment nine days before Christmas, 1965, her air wing had flown more than 12,000 sorties and dropped in excess of 10,000 tons of bombs. *The Mighty O*'s first combat deployment since the Korean War also resulted in the loss of 12 airmen—seven killed in action and five captured—and 22 aircraft destroyed.

Chapter 3

Oriskany crew and family members were still grieving the loss of loved ones from the last Vietnam deployment when, on May 26, 1966, *Oriskany* and Carrier Air Wing-16, sailed for Southeast Asia and a second consecutive Vietnam combat deployment.

U.S. Navy Photo

The Mighty O *departs San Diego for Southeast Asia and her second Vietnam combat deployment.*

Arriving on Yankee Station a month later, *Oriskany*'s air wing took up where it had left off nearly six months earlier, striking roads and bridges in the North and along the Ho Chi Min Trail. Sixteen days after arriving on station, *The Mighty O* lost the first of 25 aircraft she would lose during this deployment. Fifteen of the downed aviators were recovered and returned to *Oriskany*, six were killed and four taken prisoner.

After nearly four months of round-the-clock operations, the ship's company and aircrews were war weary and exhausted. Under such conditions a moment of inattention is all that was required for a disastrous accident to occur. That moment arrivedat 7 a.m., October 26, 1966.

A moment of inattention by war weary sailors and pilots proved to be a recipe for disaster for Oriskany *and her air wing during the 1966 combat deployment.*

U.S. Navy Photo
By PHC Art Giberson

The darkest day of *Oriskany*'s second combat deployment in Vietnam, and perhaps the darkest in the ship's history, began rather routinely with flight quarters around 10:30 p.m., October 25. Throughout the ship technicians, under the watchful eyes of their chiefs and leading petty officers, feverishly worked to ensure that all aircraft were fueled, armed and electronics and engine systems checked, in preparation for a midnight launch. The flight deck was a beehive of activity. Aircraft handlers positioned their birds while ordnancemen went about their duties of loading bombs, rockets, ammunition and parachute flares to illuminate the targets.

With the midnight launch complete and all sorties en route to their assigned target areas, the cycle was repeated in preparation for a second launch. As the last aircraft from the second launch was leaving the deck, first-launch aircraft were being recovered, refueled and rearmed for the next strike.

With the approach of daylight ordnance loads had to be modified. Unused target illuminating parachute flares were removed from returning aircraft and stowed below the flight deck in special pyrotechnics storage lockers. Because of the fast pace of near continuous flight operations, it was often necessary for munitions unloaded from returning aircraft to be placed on skids and moved to a secure location near the storage lockers until after the next launch. That's what happened on that fateful October day in 1966 as *Oriskany* prepared for yet another air strike against North Vietnam.

Around 6 a.m., while preparing for an early morning launch, two teenage sailors were returning Mk-24 parachute flares to their storage locker in hangar bay Number One, when one of the sailors accidentally pulled a flare ignition lanyard. Startled by a sudden hissing sound, one of the sailors, believing that the lack of oxygen in the storage locker would extinguish the flare, tossed it into the locker and slammed the door.

Made of magnesium and sodium nitrate, Mk-24 parachute flares provide illumination of two-million candlepower which can last for three to four minutes when ejected from an airplane or helicopter. When ignited, the magnesium and sodium nitrate reach temperatures of nearly 5,000 degrees Fahrenheit, making the flare nearly impossible to extinguish. Within minutes of the sailor tossing the flare into the locker, other flares inside the locker ignited, blowing the storage locker door off and turning *Oriskany* into a floating inferno. Flames and choking fumes carried by the ship's ventilation system, quickly spread throughout the ship.

Captain John Iarrobino, *Oriskany*'s commanding officer, was in his sea cabin when the duty officer called at approximately 7:20 a.m. notifying him of a fire on the hangar deck. The most tragic day of the captain's career had begun.

Within seconds of receiving the call, Iarrobino rushed to the bridge where he saw dense smoke engulfing the entire forward part of the flight deck and island.

Oriskany immediately went to General Quarters as the skipper maneuvered the ship allowing the wind to blow smoke from starboard to port making it easier for firefighters to get at the fire.

As fire crews fought to control the fire Captain Iarrobino noticed flames coming from the "bomb farm," an area on the starboard sponson were bombs were temporarily stored for easy access. There was no doubt in Iarrobino's mind that if the bombs and other munitions stored there exploded the results would be catastrophic.

But to his amazement, and pride, the captain saw the *Oriskany* men of all ranks, manhandling bombs over the side. He later wrote that their quick reaction to a very serious threat is what saved *The Mighty O.*

U.S. Navy Photo

The quick response of Oriskanymen in jettisoning bombs and other forms of ammunition over the side possibly saved The Mighty O *from a catastrophic disaster.*

The first casualty report was received on the bridge at around 8:50: Lt. Cmdr. William Garrity, *Oriskany*'s Catholic chaplain had been found in a passageway just forward of the hangar deck. He had obviously been attempting to make it topside when he succumbed to the smoke and flames.

Fortunately *Oriskany* was not alone in the Tonkin Gulf. The carriers Constellation (CVA-64) and Roosevelt (CVA-42) were also on station and quickly steamed to *The Mighty O*'s aid with fire-fighting equipment, doctors and chaplains.

Famed comedian Martha Raye, who was also a registered nurse, was onboard Constellation as part of a USO troupe and offered to assist the *Oriskany* medical staff in caring for the injured. Her offer was declined because Captain Iarrobino felt that under the circumstances *Oriskany* was a far too dangerous place for the famous entertainer.

Throughout the ship, people were dying and others were trapped. There was nothing they could do but wait for the fires to be brought under control so rescue parties could reach them.

Captain Iarrobino slowly began to receive progress reports. Fires were under control and dozens of men had been rescued. Unfortunately, the reports also brought more tragic news of casualties. When the final report was in Captain Iarrobino stared in horror and disbelieve at the figures: Forty-four men—36 officers and eight enlisted men had died as a result of the fire.

Three of the enlisted, two boatswains mates and a seaman, died in a passageway immediately forward of the number one elevator, at their general quarters station. The other five were trapped by fire, smoke and flood water in the ship's public affairs office on the third deck. Twenty-four of the officers were pilots who only a few hours before were flying combat missions over North Vietnam. Some were badly burned and others looked like they were sleeping—they died of asphyxiation from the heavy, dense smoke that invaded their living quarters.

U.S. Navy Photo

Smoke billows from The Mighty O *after an Mk-24 aircraft parachute flare ignited in a flare storage locker on October 26, 1966. The fires caused extensive damage to* Oriskany *and her aircraft and killed 44 sailors.*

The high proportion of officer deaths, according to after action reports, resulted from the proximity of the flare storage locker to the forward officers' country where most of the junior officers' staterooms were located. Fire, intense heat and dense, acrid smoke had raged through the staterooms and passageways arranged in a four-level, U-shaped cluster, around the number one aircraft elevator well. Many individuals died as they lay sleeping. Others were awake and tried to make their way to safety and became disoriented in the smoke and darkness when power to the forward part of the ship was lost. With their escape routes blocked, they became trapped and soon their oxygen supply was totally exhausted.

When the fires were extinguished the damage became all too visible. Two helicopters were destroyed, four Skyhawks badly damaged, hangar bay one was gutted, the catapults were out of commission and officers' country destroyed.

Departing the gun-line *Oriskany* sailed for Subic Bay to make temporary repairs and transfer the bodies to be flown home. En route to San Diego all hands assembled on *The Mighty O*'s flight deck for a memorial service and a final farewell to their 44 ship and squadron mates who had perished in the fire, as her destroyers escorts steamed alongside. This wasn't just an *Oriskany* tragedy but a Seventh Fleet tragedy, one that affected every sailor, soldier, marine and airman serving in Southeast Asia.

U.S. Navy Photo

A-4 Skyhawks were moved to Oriskany's *flight deck in an effort to protect them from the raging fires below in the hangar deck.*

U.S. Navy Photo by JOC Dick Wood

Seventh Fleet destroyers join Mighty O *crewmembers in an at sea Memorial Service for the 44 men who lost their lives in the October 26, 1966 fire.*

Chapter 4

After shipyard workers at Hunter's Point Naval Shipyard at San Francisco repaired the fire damage, *Oriskany* returned to San Diego to allow the crew to spend some highly valued time with their families before again putting to sea for carrier qualifications and workups in preparations for her third Vietnam combat deployment.

The 1967 deployment would be the last for *The Mighty O*'s workhorse air wing, CVW-16. Carrier Air Wing-16 had deployed aboard *Oriskany* for each of the carrier's previous Vietnam combat deployments and had suffered substantial losses: 47 aircraft and 20 airmen captured or missing in action. The wing's heaviest toll, however, was yet to come. Despite its losses, CVW-16 had left its mark on Southeast Asia by flying 9,551 combat sorties and dropping more than 7,000 tons of bombs on North Vietnam.

The first loss of the 1967 deployment occurred within a few hours of arriving on Yankee station. During the next seven months *Oriskany* and CVW- 16 would lose a total of 39 aircraft. Seventeen airmen were killed; two were listed as missing in action (MIA) and six, including U.S. Senator (then Lt. Cmdr) John S. McCain III (who had recently transferred to *Oriskany* from the *Forrestal*), were captured. The only bright spot of the deployment occurred on December 14, 1967, when Lt. Richard W. Wyman, an F- 8E pilot, shot down a MiG- 17—the second downed by CVW- 16 during its three deployments aboard *Oriskany*.

CVW-16's final deployment aboard *The Mighty O* brought about a slight change in one of its squadron's aircraft composition: Heavy Attack Squadron-Four (VAH-4).

Although a VAH-4 detachment had deployed aboard *Oriskany* on her previous deployment, this time the squadron flew a modified version of the A-3 Skywarrior. Nicknamed the "Whale," A-3s were the largest and heaviest aircraft designed for routine operation from aircraft carriers.

U.S. Navy Photo

Detachments of KA-3Bs and EKA-3Bs regularly flew from carriers operating in the Gulf of Tonkin. By being able to refuel aircraft that had suffered damage to their fuel systems, KA-3Bs permitted the damaged aircraft to reach their carriers and were responsible for saving many lives The EKA-3Bs assisted by providing vital intelligence on the North Vietnamese radar system and escorted most strikes, jamming enemy radar and communications.

Though heavy, 82,000 pounds, the A-3 was a highly adaptable aircraft and had several sub-variants: The EA-3B was used as electronic countermeasures aircraft, the KA-3B was an in- flight tanker version, the EKA-3B served as a multi-role tanker/ ECM/strike aircraft, while the RA-3B was equipped as a photo- reconnaissance aircraft. Still a fourth version, the VA-3B was used as a VIP transport. The Commander-in-Chief, Pacific Fleet, based at Pearl Harbor, Hawaii, routinely flew aboard the VA-3B when visiting ships in the Tonkin Gulf.

While each of the various versions played a crucial role during the Vietnam War, for pilots and air crews struggling to get battle damaged, low on fuel aircraft, back to their carriers and bases, none provided a more valuable service than the KA-3B.

Oriskany was only 13 days into her third Vietnam deployment when she received an all too familiar distress signal. "Fire on the flight deck! Fire on the flight deck!" This time, however the call came not from her own public address system, but from a sister carrier—USS *Forrestal* (CVA-59).

Operating within the vicinity of *Oriskany*, *Forrestal* had been launching air strikes against targets in North Vietnam for four days. Her crew was well trained and launches and recoveries were being carried out without the slightest hitch. Then, around 10:50 a.m. tragedy struck. A Zuni rocket, accidentally fired by an electrical power surge from an F-4 Phantom streaked across the deck and struck a 400 gallon belly fuel tank of an A-4D Skyhawk getting ready to launch.

The Skyhawk's pilot, future presidential candidate Lt. Cmdr. John McCain, managed to escape from his plane by climbing out of the cockpit, walking down to the nose of the plane and jumping off the refueling probe.

Forrestal wasn't as lucky. The ruptured tank of the Skyhawk spewed highly flammable JP-5 fuel onto the deck, instantly igniting and spreading flames over the flight deck and under other fueled and armed aircraft ready for launch. The ensuing fire caused ordinance to explode and other rockets to ignite. Flames, spread by the wind, quickly engulfed the after section of the ship turning the flight deck into a blazing inferno.

Other Seventh Fleet ships operating in the area, including *Oriskany*, immediately rushed to the assistance of the stricken carrier, just as they had done when *The Mighty O* had been in danger in these same waters a year earlier. For several hours *Oriskany* helicopters ferried firefighting equipment, firefighters and medical supplies and personnel to the fire-ravaged *Forrestal*.

It took 24 hours to contain the *Forrestal* inferno and by the time it was all over, 134 men had lost their lives, hundreds more were injured and more than 20 planes had been destroyed. It was the worst non-combat-related accident in American Naval history.

The *Oriskany* fire, when compared to *Forrestal*, is considered by many senior Navy damage control experts as a minor disaster. Though there is no such thing as a "minor disaster" aboard Navy ships at sea, the *Forrestal* fire did lay the ground work for shipboard fire fighting training that is still in use today.

After two deadly fires in as many years aboard the Navy's premier warships, virtually all new Navy recruits were required to view a training video titled "Trial by Fire" produced from factual footage of the *Forrestal* fire and the damage control efforts, successful and unsuccessful, that were used by firefighting and damage control parties aboard *Forrestal*.

U.S. Navy Photo

Forrestal *crew members fight a series of fires and explosions on the* Forrestal's *flight deck, in the Gulf of Tonkin, July 29, 1967.*

Renewed emphasis on fire fighting and damage control techniques weren't limited to recruits, however. Beginning in the late 1960s special training in shipboard fire fighting techniques was required for the crews of all deploying ships and squadrons.

As one of the fire victims, *Oriskany* took the revised training a step further and created special Fire Emergency Response Teams and conducted daily drills. Receiving highly specialized training in all type of fire fighting techniques, the Fire Emergency Response Teams were the first responders for each and every fire, regardless of size or class. When the fire alarm sounded aboard *Oriskany*, day or night, the teams were on the scene within minutes.

The Fire Emergency Response Teams took so much pride in what they had learned that after a while, each alarm, whether an actual fire or a drill, was treated as a competition among the various team members to see who would be the first on the scene. But, this wasn't a game. The competitive spirit of these fire fighting and damage control teams was truly a matter of life and death.

After the deadly fire that claimed the lives of 44 Oriskany *sailors in 1966, daily fire drills under the supervision of special Fire Emergency Response Teams trained in shipboard emergency fire fighting techniques, ensured that Oriskany sailors fighting the war in Vietnam, lived to fight another day.*

U.S. Navy Photos
by PHC Art Giberson

Having done all she could to assist her fellow carrier; *Oriskany* returned to her position on the gun-line and resumed launching attacks against North Vietnamese targets which had grown far more hazardous than the attacks CVW- 16 had conducted on previous deployments.

There were two major reasons why *Oriskany*'s third Vietnam combat deployment was more costly in terms of pilots lost: Since her last deployment the number of SAM sites (surface-to-air-missile) ringing the Hanoi, Haiphong and other industrial and military areas had greatly increased.

Pilots were finally given the go-ahead to attack targets in the heart of North Vietnam which had previously been off limits to American air strikes: the Lach Tray Shipyard at Haiphong, port facilities at Cam Pha and the huge Phuc Yen MiG Base just north of Hanoi.

The lifting of target restrictions and the increased number of SAM sites made for hazardous flying as *Oriskany* pilots unfortunately discovered during an August 31 attack on the port city of Haiphong. The air wing had been conducting strikes on Haiphong for two consecutive days. On the third day as a flight of 10 aircraft approached Haiphong air space, the flight leader spotted two SAM missiles lifting off from the Haiphong area and alerted the pilots to commence evasive maneuvers in an attempt to evade the incoming missiles.

Most of the *Oriskany* birds managed to dodge the "flying telephone poles," as the SAM's were dubbed by air crews flying bombing missions over North Vietnam. Three however, found their targets.

Losing three aircraft in a single raid was a major blow to CVW-16 and indeed the entire ship. Of the three downed pilots, two: Lt. Cmdr. Hugh Stafford and Lt. Jg. David Cary safely ejected from their aircraft and were captured. The third, Lt. Cmdr. Richard Perry was killed.

The loss of Stafford, Perry and Cary wasn't in vain, however. Their fellow CVW-16 pilots took out a major railroad/highway bridge in the center of Haiphong that had previously been on the "Off Limits" list.

Oriskany was nearing the final 90 days of her third Vietnam deployment, when Lt. Cmdr. John McCain was launched from the deck of *The Mighty O* for his 23rd bombing mission over North Vietnam. As the sleek A-4 Skyhawk left *The Mighty O's* deck that October day, it probably never entered his mind that he too, would soon become a victim of the deadly SAM's.

A 1958 graduate of the U.S. Naval Academy, McCain had volunteered for duty with Attack Squadron-163 onboard *Oriskany* after the *Forrestal* incident. McCain broke both arms and a leg during ejection from his A-4 Skyhawk after being hit by a surface to-air-missile. Captured upon landing he was beaten by his captors until he was very near death. Years later, after being released from captivity a reporter once referred to McCain as a hero, to which he reportedly quipped, "Do not call me a 'war hero'... I am anything but! The fact that I was incompetent enough to get shot down should dispel any notion that I am a hero."

Hero or not, John McCain was held prisoner for five-and-a-half-years, mostly in the infamous Hanoi Hilton. When the North Vietnamese discovered that McCain was the son of the Commander-in-Chief, U.S Pacific Fleet, he was offered a chance to go home, in an effort to embarrass the U.S. Government. McCain refused to be repatriated and remained with his fellow POWs until they were all able to come home.

U.S. Navy Photo

Lt. Cmdr John McCain leaves the aircraft that brought him and other POWs from North Vietnam to Clark Air Force Base on Philippines after his releas e in 1973.

Following his release in 1973, McCain was reinstated to flight status and became commanding officer of Attack Squadron-174. Later he became the Navy liaison to the U.S. Senate. McCain retired from the Navy as a captain in 1981.

The following year he was elected to Congress representing what was then the first congressional district of Arizona. In 1986, he was elected to the United States Senate.

The lifting of target restrictions, which many senior Defense Department officials had been demanding, not only resulted in more American aircraft and air crews being shot down it also inflicted far greater damage on a determined enemy. But for Lt. Richard Wyman, a fighter pilot with VF-162 the lifting of restrictions brought an early Christmas gift.

While flying a combat air support mission over North Vietnam on December 14, 1967, Lt. Wyman engaged a MiG-17, and for more than 15 minutes, Wyman's F-8E Crusader played cat and mouse with the MiG-17 in a furious dogfight which ranged far and wide over North Vietnamese airspace The aerial battle ended when Wyman fired an air-to-air missile, locking on to the MiG-17 and sending it flaming into a rice paddy where it exploded and burned.

On the previous deployment, another VF-162 fighter pilot, Cmdr. Richard M. Bellinger recorded *Oriskany*'s first kill of the Vietnam War when he too, splashed a MiG-17 in an aerial gun battle.

The Mighty O lost another pilot just eight days before ending her most deadly deployment of the war. On January 5, 1968, Lt. Ralph Eugene "Skip" Foulks Jr., of VA-163 was killed while flying one of the final combat missions of the deployment. Foulks' remains were later recovered and returned to his family in March 1993—25 years after his A-4E Skyhawk was shot down over North Vietnam.

After 122 days on the gun-line, *Oriskany* turned her bow westward and sailed for NAS Alameda, her new homeport.

In a letter to his mother, Elvah Jones, a resident of Pensacola, Florida, the young naval aviator expressing his love of flying, wrote: "Mom, it's so peaceful and calm up here that you almost want to stay forever. It's as if you could reach out and touch the face of Jesus."

During her third Vietnam combat deployment, *The Mighty O* had set the operational standard for other carriers and air groups to follow for the reminder of the war.

Of *The Mighty O's* deployment's, this was the most spectacular in her quarter century of service. Her pilots flew an unprecedented 181 air strikes in the highly SAM infested northeast section of North Vietnam which included the cities of Hanoi and Haiphong.

In addition to air strikes on the Hanoi and Haiphong areas, CVW16 air crews flew a total of 9,552 combat and combat-supported missions dropping more than 7,500 tons of ordnance on enemy targets.

Tragically, *The Mighty O* lost 13 airmen in that deployment. Six of the 13 are still unaccounted for. For their dedication and airmanship during her third Vietnam deployment, *Oriskany* pilots were awarded a total of 1,411 medals: four Navy Crosses, six Silver Stars, two Legions of Merit, 96 Distinguished Flying

Crosses, six Bronze Stars, 1,978 Air Medals, 192 Navy Commendation Medals, 127 Navy Achievement Medals and 10 Purple Hearts. Sixty-five Commander-in-Chief Pacific. Letters of Commendation and 77 Commander Seventh Fleet Letters of Commendation were also awarded. The ship itself was awarded two Navy Unit Commendations.

U.S. Navy Photo

A war weary Oriskany *arrives at her new homeport, NAS Alameda, Oakland, California, January 31, 1968.*

Returning to the United States on January 23, 1968, the *Oriskany* shifted to her new homeport at NAS Alameda, Oakland, California. Her stay at the third, and final, homeport was short, but highly welcomed. To show appreciation for what *Oriskany* and her air wing accomplished the Oakland City Council, despite near daily war protests being conducted across the bay in San Francisco, in a special October 24, 1968 City Council meeting adopted the grand old warrior as their own. Oakland government officials and its citizens were proud that NAS Alameda would be *The Mighty O*'s homeport and they made certain that it was a well documented fact.

Four months later, February 8, 1968, *Oriskany* cast off her lines, steamed across the bay and once again entered Hunters Point Naval Shipyard where she received an extensive nine-month overhaul and maintenance period. The extended yard period ensured *The Mighty O* would be able to return to Vietnam for four more combat deployments.

In early 1969, shipyard work complete, *Oriskany* took on provisions and munitions and put to sea for refresher training and flight qualifications with her new air group, Carrier Air Wing-19 in preparation for her fourth Vietnam combat deployment.

U.S. Navy Photo by PHC Art Giberson

Like a fish out of water, The Mighty O *sets in dry dock at Hunter's Point Naval Shipyard. After a year-long yard period* Oriskany *returned to the Western Pacific for her fourth combat deployment.*

Chapter 5

Final pre-deployment operations complete, sailors, marines and air wing personnel, attempting to prolong their goodbyes as long as possible, lingered on the pier with family members, friends and well-wishers as final preparations were made to get underway. Then, approximately 10:15 a.m., Wednesday, April 16, 1969, lines were cast off, colors shifted and *The Mighty O* was underway for still another tour of duty in Southeast Asia.

As the Golden Gate Bridge slowly faded into the distance, dozens of sailors, veterans and rookies alike (including the author), loitered about the flight deck; lost in their own thoughts about what lay ahead. *Oriskany* was returning to war and they knew, not everyone would be returning.

These first few hours at sea, after nearly a year at Hunter's Point, were a time of readjustment. A time to reestablish sea legs and get used to the sounds and smells of the ship: waves slapping against the hull as *The Mighty O* plowed through the water, wind and salt spray whipping across the flight deck, the fluttering of dozens of multi-colored signal flags flying from the rigging. The odor of jet fuel, salt air, dozens of varieties of after shave, fresh brewed coffee, food being prepared, and countless other unidentified odors, filled their nostrils as they grabbed one final look at the rapidly disappearing California coastline.

After brief stops at Hawaii and Subic Bay, *Oriskany* took up position on Yankee Station on May 16. Fortunately, during the final deployment of the decade, the air war had subsided somewhat. Although *Oriskany* was on the gun-line for 116 days, her losses were relatively light compared to her previous deployment. Nevertheless, 10 aircraft were lost and three men—two CVW-19 airmen and an *Oriskany* enlisted crewmember—were killed.

U.S. Navy Photo

The Mighty O *with Carrier Air Wing-19 embarked arrives on Yankee Station.*

With bombing missions, refueling escort destroyers and keeping the crew fed, the replenishment of her own supplies and refilling her fuel tanks every few days was an operational necessity. Underway replenishment was an all-hands task akin to weekly family shopping trips back home. No one looked forward to them, but it was something that had to be done.

U.S. Navy Photo

Underway replenishment generally involved two or more ships steaming alongside a supply vessel taking on everything from bullets to black oil. This re-supplying technique was the only way Oriskany *and other warships were able to remain on station and continue conducting combat operations for months on end.*

U.S. Navy Photos by PHC Art Giberson

The at sea shopping days, required all hands not actually on watch. Supplies deposited on the flight deck were quickly recovered and taken below decks. Underway replenishment also presented good photo ops and was the topic of a 1969 Department of Defense film for various media outlets about the Navy's role in Vietnam.

Former NATTU Instructor, Giberson Shares Oriskany's Underway 'Shopping'

Replenishment cycles were nearly always the same. Take on fresh supplies and fuel and back to the gun-line and the resumption of air strikes. Occasionally, however, if the crew was lucky, the ship was permitted a departure from Yankee Station for a few days R&R at Subic Bay, Japan or Hong Kong.

U.S. Navy Photo

Oriskany *at Naval Station Subic Bay, Philippines 1969, after completing a 30-day combat period on Yankee Station, in the South China Sea.*

R&R periods, whether in Subic Bay, Hong Kong or Sasebo, Japan, were highly cherished but all too brief. Still, the brief interludes allowed Oriskanymen a momentarily escape from the rigors of war. Though they may have tried, they could never totally escape the fact that while they were away from the war zone, even for a brief period, other Americans were dying and by war's end the total would exceed 58,000. They were acutely aware that after a week, sometimes two, *Oriskany* would up anchor and return to Yankee Station and the war.

En route to Yankee Station from R&R was the time to prepare aircraft and bombs for strikes on the enemy.

After seven months of combat operations, *Oriskany* returned to Alameda on November 10, 1969. During the holiday period the majority of the crew went on holiday leave, a large number were transferred or discharged, and still others temporarily departed the ship for specialized training.

With the dawning of a new decade, it was back to work, preparing for still another combat deployment. Although pre-deployment workups were routine for most of the crew, this one was slightly different because the ship's company and aircrews were preparing to go war with new aircraft. *Oriskany*'s three combat-proven A-4 Skyhawk squadrons—VA-23, VA-193 and VA-195—were replaced with two A-7 Corsair II squadrons—VA-153 and VA-155.

An Oriskany A-7 Corsair over flies The Mighty O *during Operations in Southwest Asia in 1970*

U.S Navy Photo

With two new attack squadrons on board, The Mighty O departed Alameda May 14, 1970 for her fifth Vietnam combat deployment. By the time Oriskany reached the war zone on June 14, the combat area of operations had changed. Navy, Marine Corps and Air Force aircraft were now routinely striking targets in Cambodia and Laos as well as North and South Vietnam. Eleven days after arriving on Yankee Station, The Mighty O suffered its first casualty when an A-7A from Attack Squadron-153 was lost while flying a combat mission over Laos.

Completing her first line period of the 1970 deployment on June 29, *Oriskany* sailed for Hong Kong and some much needed R&R.

U.S. Navy Photo by PHC Art Giberson

The Mighty O *at anchor in Hong Kong Harbor for a 10 day R&R. After leaving the (then) British Crown Colony,* Oriskany *and her air wing returned to Vietnam and resume d air strikes.*

While all liberty ports during the Vietnam War were, for the most part, considered good, *Oriskany* sailors were especially fond of Hong Kong because of its wide variety of recreational activities and shopping opportunities. Few service members visiting Hong Kong ever left without the purchase of a specially-custom tailored suit. For sailors and marines more interested in seeing the sights than shopping or partying, Hong Kong had no equal. This was truly a sailor's paradise.

U.S. Navy Photos by PHC Art Giberson

Oriskany *sailors disembark from a Hong Kong water taxi for a day on the town. Below, an* Oriskany *sailor photographs an old man and his granddaughter enjoying ice cream.*

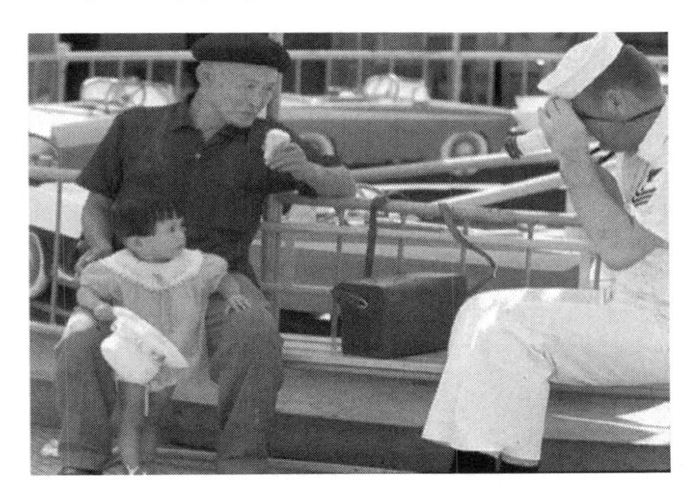

After 10 days in Hong Kong, *The Mighty O* hoisted anchor, sailed out of Hong Kong Harbor and returned to the war.

An F-8 Crusader into launch position aboard Oriskany.

U.S. Navy Photo by PHC Art Giberson

Flight operations and the multitude of other tasks a carrier normally provide for smaller task force ships such as providing chaplain support to ships too small for a permanently assigned chaplain. Ferrying men of the cloth between the various ships of the task force was the job of the "Holy Helo."

The chaplains, regardless of denomination were airlifted from *Oriskany* and lowered by sling to the deck of a pitching ship below. The story of these "Circuit Riding Chaplains" and the religious services and guidance they provided to the men of the Navy's smaller men-of-war were the subject of a full page story in the August 13, 1970 edition the armed forces daily newspaper, *Stars & Stripes.*

Without the "Circuit Riding Chaplains," and the helicopter crews who ferried them from ship to ship, as the story pointed out, organized religious services for hundreds of American fighting men in Vietnam would have been denied.

Pacific Stars & Stripes

FEATURES

OPINION, ENTERTAINMENT, COMICS, PAGES 9-16

'Holy Helo!' The Chaplain's Here

Story by
CPO ART GIBERSON

Photos by
PO. 1.C. RUSS FUHRMAN

ABOARD THE USS ORISKANY — [illegible]

By the time *Oriskany* returned home on November 29, 1970, she had been on the gun-line for 90 days; lost three aircraft and two pilots, expended nearly 6,000 tons of ordnance, and observed her 20th birthday.

U.S. Navy Photo

The Mighty O *returns from her fifth Vietnam combat deployment, four days after Thanksgiving, 1970.*

Five months later, June 4, 1971, *The Mighty O* again sailed under the Golden Gate Bridge, en route to her sixth Vietnam deployment. Fortunately *Oriskany* suffered no direct combat losses during this deployment. She did, however, lose four aircraft in operational mishaps which resulted in the deaths of two of the pilots. She returned to the States in December 1971 after only 75 days of combat.

The queen of the Essex-class carriers left the United States again on June 5, 1972, for her seventh and final deployment of the war. She was on Yankee Station for 169 days—the longest gun-line period of any of her previous Vietnam deployments.

Despite the fact that *The Mighty O* was growing obsolete after nearly a decade of combat , she was still a key factor in the infamous "Christmas Bombings" of 1972, in which American aircraft conducted some of the heaviest bombing raids of the war against Hanoi and other key North Vietnam cities. On January 27, 1973, Cmdr. Denis R. Wiechman, a VA-153 A7A Corsair pilot, flew *Oriskany*'s last combat mission of the war and his, 612th combat mission.

It was befitting that the carrier which had been the tip of the spear for the U.S. Marines when they landed at Da Nang, eight years earlier, concluded her wartime service with the longest combat deployment of her active service.

By the time she returned to Alameda on March 30, 1973, *The Mighty O* had been deployed for 10 months. During the deployment she lost six aircraft. One pilot was captured and two killed. The other three plots were rescued and returned to *Oriskany.*

During her seven Vietnam combat deployments, 81 of the ship's sailors lost their lives, including the 44 killed in the 1966 fire. Eighteen fliers were captured; three of whom died in captivity; and five were listed as missing in action (MIA). In addition to the human toll, 109 aircraft were lost. Following the war's end, *Oriskany* returned to the Western Pacific twice more before finally being decommissioned in 1976.

The Vietnam War will most likely be remembered by historians as a ground and air war, and rightly so. Hopefully they will also remember that much of that air power was made possible by *Oriskany* and the thousands of sailors who were responsible for operating the ship and launching and recovering *The Mighty O*'s war birds, and the fact that she was a founding member of the "Tonkin Gulf Yacht Club."

U.S. Navy Photo

When The Mighty O *sailed into retirement on September 30, 1976, her legacy as the Navy's only Super Essex-Class carrier had been well established. The USS* Oriskany *was, and always will be, in a class of its own.* The Mighty O *was one of a kind and remains so as the undisputed queen of Gulf Coast fishing reefs.*

After a quarter of a century of service, including nearly a decade at war, *Oriskany* was decommissioned September 30, 1976 and placed in the Reserve Fleet at Bremerton, Washington. During the early 1980's, several attempts were made to reactivate *Oriskany* and return her to the fleet, but each of the various proposals failed due to high cost and the lack of a suitable air wing.

In July 1989 *The Mighty O* was stricken from the Naval Vessel Register and sold for scrapping. The contractor defaulted however, and the Navy repossessed the rusting former hulk of an aircraft carrier.

Thoughts of scrapping or scuttling *Oriskany* may have been avoided entirely if plans for procurement of the ship for use as a tourist attraction to be called the "City of America" hadn't been scuttled in the early 1 990s. The plan called for *Oriskany* to be refurbished and anchored in Tokyo Bay as a cultural exhibit to expose the Japanese to the America lifestyle. Loud angry protests from former *Oriskany* sailors and veterans' organizations quickly scuttled the plan. In the opinion of many of the former *Oriskany*men, it would be far more dignified to allow *Oriskany* to rust and later be cut up for scrap rather than allow her to become a tourist attraction for a foreign nation.

Although one of America's strongest allies since the beginning of the Cold War, Japan was still viewed as a former enemy by many World War II veterans who felt that it would be a slap in the face to have *Oriskany* fall into the hands of the Japanese.

There was talk of turning *Oriskany* into a floating museum, where she could be visited by millions of people. Again, the cost of such a venture was out of reach and the idea was soon abandoned.

After spending several years at the former Mare Island Navy Yard, Vallejo, California, *Oriskany* was towed to the Beaumont Reserve Fleet in Beaumont, Texas, to be scrapped. But in keeping with *The Mighty O*'s fighting spirit, the plans were changed yet again and in 2004 *Oriskany* was transferred to the State of Florida to be sunk as an artificial reef.

In keeping with her long tradition of "firsts" *Oriskany* is the first warship ever to be slated for that purpose. Initially scheduled to be sunk in 2005, 22 miles off the coast of Pensacola, Florida,—the "Cradle of Naval Aviation"— where nearly all of the aircrews who flew from *Oriskany* received their wings.

Before scuttling the ship a detailed environmental assessment was required to determine what, if any environmental impact the sinking might have on the surrounding waters. After an extensive study, the green light for the project was given by the Naval Sea Systems Command in December 2004 and the carrier was towed to Pensacola from Texas to undergo final preparations for the eventual sinking.

Docked for several months at the Port of Pensacola, the once "*Mighty O*" became an instant "must see" for tourists, former *Oriskany* crewmembers, and air wing personnel residing in the Florida Panhandle and the Alabama and Mississippi Gulf Coast. As the planned sinking date neared, thousands of people from around the nation, including more than 600 former *Oriskany* crewmembers, planned trips to Pensacola in hopes of witnessing the *Oriskany* go to her final resting place.

Those plans, as well as the planned June 25, 2005 sinking of *Oriskany* were stopped cold when Hurricane Ivan struck the Pensacola area on September 16, 2004. Although

The Mighty O easily survived Ivan's fury, environmental concerns forced the postponement of several planned sinking dates and in June 2005, *Oriskany* was towed back to Texas to ride out the 2005 hurricane season.

Photo by Art Giberson

Russell Furiman, a former Oriskany *crewmember is reunited with his old ship after more than 30 years. Furiman, a resident of Ogden, Utah, was visiting Pensacola in 2005, when he learned* Oriskany *was at the Port of Pensacola, waiting to be towed into the Gulf of Mexico where it was to be sunk. Furiman served aboard* Oriskany *as a Photographer's Mate during the Vietnam War.*

Nine months later, March 2006, *Oriskany* returned to Pensacola and temporarily docked at the Pensacola Naval Air Station while final preparations for her sinking were made. In a final salute to *The Mighty O*, a memorial service was held at the

National Museum of Naval Aviation, onboard the Pensacola Naval Air Station, May 13, 2006 to pay tribute to *Oriskany* and her crews for exemplary service to the Navy and the United States of America.

Photo by Art Giberson

Former Oriskany *crewmembers gather at the National Museum of Naval Aviation May 13, 2006, for a reception and memorial service in honor of* Oriskany *and former shipmates.*

Former Oriskany *crew member James Salisbury and his wife Mary, of Plainfield, IL, at Pensacola Naval Air Station, bid a final farewell to* The Mighty O. *May 15, 2006, as she is towed to her final destination, 22 miles offshore.*

10:25 a.m., May 17, 2006: *Thirty-six minutes after 500 pounds of explosives were detonated deep inside* Oriskany, *the gallant warship, listed to port and quietly slipped beneath the surface of the Gulf of Mexico. Two days later divers were permitted to dive on* The Mighty O, *upright in 212-feet of water.*

Epilogue

The Mighty O, her air groups having long since made their final fly away, settled into her final homeport on a white, sandy sea bed, 212-feet below the surface of the Gulf of Mexico; approximately 22 plus miles off the coast of Pensacola, Florida, in May 2006. *The Mighty O* is the first aircraft carrier intentionally scuttled to form an artificial reef, continuing a long list of *Oriskany* "firsts," began in 1951 when she became the first aircraft carrier to have a complex automatic steering system.

Oriskany will be crewed by such marine life as barnacles, sponges, plants, clams, and coral. These organisms, according to marine biologists, will form the basis for an ocean reef community that will support many fish species and provide a feeding ground for larger fish as well as an environmentally safe artificial reef suitable for diving and fishing.

While the keel of *The Mighty O* rests at 212-feet below the surface, her open hanger deck, at a depth of around 160-feet, provides a habitat for all sorts of marine life. The flight deck is 137-feet beneath the surface and easily accessible to most experienced divers. The most reachable part of the ship for divers is the top of the island at 70-feet below the surface.

The area surrounding *Oriskany's* final homeport is home to numerous fish typically found off Florida's coast: Spanish mackerel, black drum, snapper, bonito, flounder, red snapper, triggerfish and grouper, etc. Marine biologists say the area is a normal migration route for whales, dolphins and sea turtles and that *Oriskany* will eventually be colonized by a wide variety of marine invertebrates and fish.

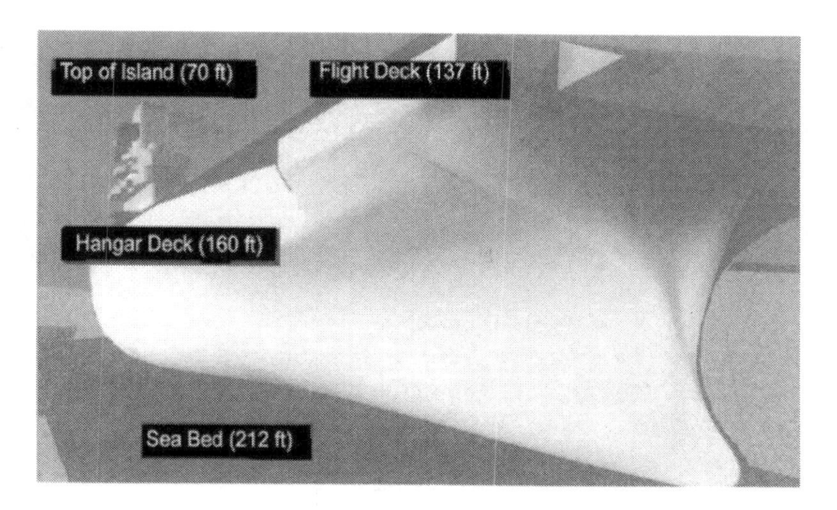

Approximate depths of Oriskany *from water surface.*

Tourism, fishing and diving experts have predicted the 911-foot aircraft carrier will become a fishing hot spot and an exceptional destination for divers year-round. With *The Mighty O* as a backdrop divers have the opportunity to swim with goliath grouper, ocean sunfish and eagle rays. From the surface, anglers will be able to reel in popular game fish such as grouper, snapper and amberjack. The thousands of sailors, marines and airmen, who served aboard one of the Navy's great warships, can relax with the knowledge that their ship, *The Mighty O*, continues its service to America.

About the Author:

Art Giberson, a retired Navy Chief Photographer's Mate made two combat deployments aboard the USS *Oriskany*—1969-1970—as OP Division Leading Chief/Division Officer.

In addition to ***The Mighty O*** he is the author of four other non-fiction books: ***WALL SOUTH***, a pictorial history of the Vietnam Veterans Memorial in Pensacola, Florida; ***The BLUE GHOST*: *The Ship That Couldn't Be Sunk***; ***EYES of the FLEET: A History of Naval Photography***; ***Century of War, War Stories*; *The Crazy Ones Shot Film*** and three novels, ***PHOTOJOURNALIST: The Story of a Navy Combat Photographer*, *Anything but the Truth***; ***Brinkmanship: Standoff in the Caribbean***. The author is a member of the USS Oriskany Association, International Combat Camera Association and the National Association of Naval Photography.

A native of Bluefield, West Virginia, the author and his wife, Jean, reside in Pensacola, Florida near their four children and 13 grandchildren.

The Mighty O Vital Statistics

USS *Oriskany* "*The Mighty O*" is 911 feet long. Eight boilers and four steam turbine engines, delivered 150,000 shaft horsepower to four propellers which drove *Oriskany* through the water at speeds in excess of 30 knots. During fleet operations *Oriskany* consumed in excess of 200 thousand gallons of fuel oil per day.

Comparable in height to a 25-story skyscraper, *The Mighty O* has 10 decks and extended 192 feet above the waterline. *Oriskany*'s crew, including the air wing, consisted of around 3,500 men. During the Vietnam War, *Oriskany* carried a complement of 80 aircraft (jets, prop-driven and helicopters). Most of her fixed-wing aircraft, including the 82,000 pound A3 Sky Warrior, were launched by two steam-driven catapults and recovered by four arresting cables.

During the Korean and Vietnam Wars *The Mighty O* was on the gun-line for a total of 922 days and lost 106 aircraft to combat and other operational causes.

34347442R00042

Made in the USA
Lexington, KY
22 March 2019